ACKNOWLEDGEMENTS

The lessons and ideas shared in this book are a compilation of several years of work with students in schools and school-counselors-in-training.

This book is dedicated to all the children who believe in the best the world has to offer even when what happens in the world doesn't make sense. It is also dedicated to school counselors who continue to believe in the best in students and help students feel hopeful about the future. The best is yet to come.

FOSTERING RESILIENCE AND STRENGTH

TABLE OF CONTENTS

INTRODUCTION

The activities in this book were developed with the concepts of resilience and positive psychology in mind. Research indicates that characteristics of resilient youth include an internal locus of control and a connection with at least one significant adult in their lives. Additionally, positive psychology, developed by Martin Seligman, Ph.D., suggests we focus on "signature strengths" or traits that will help us have a positive approach toward life and will help students think positive thoughts instead of negative ones (Seligman, 2002). In addition, positive psychology is the study of what gives our lives meaning, and there are protective factors (both internal and external) that promote psychosocial resilience in children and youth (Masten, Cutuli, Herbers, & Reed, 2009).

The connection between resiliency and a strengths-based view of the world is documented in the literature. Bosworth and Walz (2005) indicated that students need a positive and protective environment that includes natural and caring interactions between students and school personnel. Further, they note that it is essential to accept students as they are and that the focus should be on whatever is good about each individual student. These concepts promote environmental caring and student strengths, which are essential to resiliency.

The book is divided into two sections: resilience building and strengths building. Each lesson is aligned with the ASCA Mindsets & Behaviors and should align with state standards. It should be noted that the lessons do not have to be used in sequence, but you should begin and end with the lifeline activity if you are using the resilience activities only. The lifeline activity is a great place to start with students and works well repeated as the last session. Focusing on the future and goal-setting of the lifeline is an effective way to bring the sessions to closure and to promote hopefulness for the future.

It is my hope that the activities provided in this resource are useful for you. Many of the activities give students the opportunity to reflect on their life and the events occurring in their world. There are activities fostering interaction and fun while others may require greater introspection and more focus on the individual. The primary goal for school counselors who use these activities is to help relate to students where they are and to be a catalyst for instilling the students' belief in their own ability to build a bright future and to face adversity with optimism and hope.

Why This Resource is Needed

It is no secret that students have seen the best and worst in the world today. Amid the excitement of new technology students have seen worldwide violence, natural disasters, school massacres and a country that has been at war for most of their lives. Despite the achievements in technology and the opportunities available, there is often a great deal of devastation and heartache, both in their personal lives and in the world around them. In addition, the pressure to achieve and excel may drive students to despair when something does not go the way they had planned. The ability to bounce back – to be resilient – should be developed along with other life skills that help students achieve success in school and in life.

While many students feel hopeful about the future and have some sense of personal success, they may also experience anxiety, fear and great uncertainty about the future, both the world's and their own. School counselors are integral change agents who are in a pivotal position to help students develop the resilience and strength to overcome difficulty as they move with hope through their school careers.

How to Best Use This Resource

The sessions/activities are designed to engage students in the counseling process and invest them in their social/emotional and academic success. In addition, the activities are intended to help students identify their strengths and their support systems so they can access help, both internally and externally, in times of need.

The target audience for these activities could be students who have already overcome great difficulty and adversity in their lives, students who might have a propensity toward difficult or adverse situations and/or students who need a boost in self-concept or self-efficacy. In fact, all students could benefit from participating in these activities. The activities are best for students in grade 4–8 due to the cognitive level of consideration and thoughtfulness some of the activities require.

In addition to using these activities individually or in a small group, many of these concepts may be used when conducting classroom lessons. Some of the activities could be easily adapted for large groups; however, one of the important components of resilience is the connection between the student and a significant adult (hopefully you, the school counselor).

When using the lessons in small-group counseling, it is often good to keep the activity pages together in a folder or blank book so students can keep their work at the end of the group as a reminder of their resilience.

The sessions are designed to be conducted in 30-45 minutes. Modifications should be made as needed to meet students' needs and academic program. The pre- and post-test assessments allow you to measure a change in perception data as a result of the activities. Consider using more objective measures of change as well, such as improved attendance, behavior or achievement.

SECTION 1
RESILIENCE-BUILDING ACTIVITIES

As today's youth see more and more tragedy and experience life's hardships, it is critical to provide them with the skills to confront hard times and to develop attitudes that foster the ability to do and be something better. The resilience-building sessions focus initially on students' experiences and help them understand that many things that happen (good and bad) are out of their control. Despite the lack of control, however, students can decide how they will respond or react to the difficulties in their life by recognizing their internal assets that help them have more control in their future. Identifying internal assets and external support systems helps students build resilient attitudes to handle life's challenges.

Pre-test/Post-test: Resilience

When something bad happens to you, name three ways you handle it.

____

____

____

When you need to talk to someone about a problem, how many people can you think of that you could talk to? (Circle one).

1–2	**3–5**	**5–10**	**10+**

On a scale of 1-5 (with five being the best, three being OK and one being the worst), how do you think you handle things when they don't go the way you would like?

1	**2**	**3**	**4**	**5**
(Terrible)	(Not OK)	(OK)	(Better than OK)	(Awesome)

On a scale of 1-5 (with five being the best, three being OK and one being the worst), how do you feel about your future right now?

1	**2**	**3**	**4**	**5**
(Terrible)	(Not OK)	(OK)	(Better than OK)	(Awesome)

On a scale of 1 to 5 (with 5 being very hopeful and 1 being not hopeful at all), how hopeful do you feel about what is happening in the world?

1	**2**	**3**	**4**	**5**
(Not hopeful at all)	(Not hopeful)	(OK)	(Somewhat hopeful)	(Very hopeful)

On a scale of 1 to 5 (with 5 being very hopeful and 1 being not hopeful at all), how hopeful do you feel about what is happening in your school?

1	**2**	**3**	**4**	**5**
(Not hopeful at all)	(Not hopeful)	(OK)	(Somewhat hopeful)	(Very hopeful)

On a scale of 1 to 5 (with 5 being very hopeful and 1 being not hopeful at all), how hopeful are you that you can control or change things that are happening?

1	**2**	**3**	**4**	**5**
(Not hopeful at all)	(Not hopeful)	(OK)	(Somewhat hopeful)	(Very hopeful)

LESSON ONE: MY LIFE STORY

Summary

Students will list and discuss significant events in their life so far and reveal who had control of that event happening and their feelings about that event. This activity allows the school counselor to get to know the students and have them share their lives in a nonthreatening way.

ASCA Mindsets & Behaviors

- M 1. Belief in development of whole self, including a healthy balance of mental, social/ emotional and physical well-being
- B-SMS 7. Demonstrate effective coping skills when faced with a problem
- B-SMS10. Demonstrate ability to manage transitions and ability to adapt to changing situations and responsibilities

Objectives

Students will:

- Identify significant life events (good or bad)
- Discover they often did not control what happened in their lives

Materials

Copies of the pre-test on resilience, "What's Your Story?" handout, blank paper for brainstorming, pen or pencil for writing

Procedure

1. Administer the pre-test on resilience.
2. On the scratch sheet of paper, have students list the significant life events that have happened from their birth to the present. Students may include any events, happy or sad.
3. Ask students to transfer this information to the "What's Your Story?" page on the side labeled "Events."
4. Once all the life events have been listed, go back and discuss how students would feel if that event occurred today, who or what controlled that event and how they felt at the time the event occurred. The goal is to have students see that some of the adverse events in their life have been out of their control.
5. Have students list their feelings and control responses on the "Feelings/Control" side of the lifeline. Allow group members to share as much as they would like. They may want to share only one event or all of them. Ask:
   - What kinds of feelings do you have when you see the things that have happed in your life so far?
   - How did you cope with (or handle) each event?
   - If there were things you could have controlled, what would you have done differently to change things?
6. Students may put their timeline in their lifebooks or folders to refer to later.

Plan for Evaluation

Administer pre-test resilience survey.

Perception Data: Collect data from pre-test survey to assess student resilience at beginning of the sessions to compare with post-test responses.

Outcome Data: Check other areas where students might need support: academic, attendance, behavior. Note student performance and behavior in any of the areas to compare with data collected after the sessions are completed.

What's Your Story?

Events

Feelings/Control

______________________'s Lifetime

LESSON TWO: WHO SURROUNDS YOU?

Summary

Students will identify the people in their lives who surround or support them and discuss how each person helps or influences their lives.

ASCA Mindsets & Behaviors

- M 5. Belief in using abilities to their fullest to achieve high-quality results and outcomes
- B-SMS 7. Demonstrate effective coping skills when faced with a problem
- B-SS 2. Create positive and supportive relationships with other students
- B-SS 3. Create relationships with adults that support success

Objectives

Students will:

- Use communication skills to know when and how to ask for help when needed
- Select resource people in the school and community and know how to seek their help

Materials

Who Surrounds You? handout, pen or pencil

Activity

1. Distribute the "Who Surrounds You?" handout.
2. Discuss what it means to have people who support you.
3. Have participants list the people they can rely on to help them when things go wrong.
4. Have participants write below each circle how each person helps them.

Processing

1. Discuss how you often rely on people for different situations because they all support you in different ways. Ask students to consider if any of these people might surround them in negative ways (e.g., may be a bad influence or pressure to make bad choices).
2. Have students share how each person who surrounds them can help or support them.
3. Tell students to use this activity to remind them who they can turn to for support.
4. Ask students to keep a copy of this sheet accessible or add it to the lifebook they are creating. Make a copy to keep in the school counseling office in the event a student needs to be reminded of the human resources and external assets available to him/her.

Relationships: Who Surrounds You?

Identify five people in your life who surround you. The middle is you.

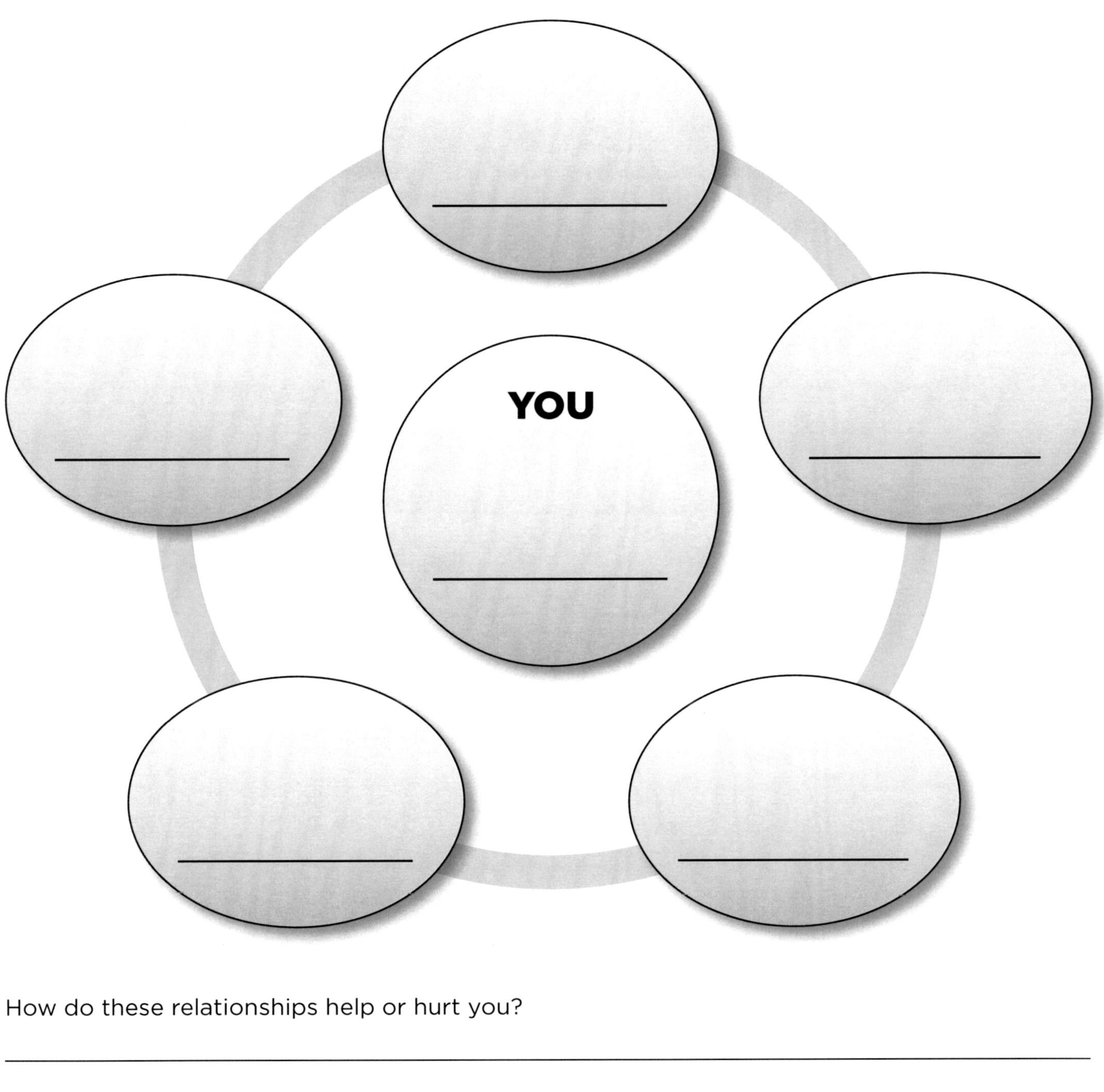

How do these relationships help or hurt you?

____

____

What can you do to maintain or improve these relationships?

____

____

If you could, would you add more circles? Turn this page over and add any more circles or people you would like to add.

LESSON THREE: MY STRENGTHS AND TALENTS

Summary

Students will identify the internal assets (strengths and talents) they possess and consider how those strengths and talents help them in their life.

ASCA Mindsets & Behaviors:

- M 5. Belief in using abilities to their fullest to achieve high-quality results and outcomes
- B-LS 1. Demonstrate critical-thinking skills to make informed decisions
- B-SMS 7. Demonstrate effective coping skills when faced with a problem

Objectives

Students will:

- Recognize and describe personal strengths and assets as skills they can use to solve problems
- Generate ways that they can use personal strengths and talents as coping strategies for daily life

Materials

Strengths & Talents handout, scratch paper, pen or pencil

Activity

1. On a scratch sheet of paper, ask the students to generate a list of strengths and talents they possess that help them get through hard times. Define the terms as follows:
   - Strength: Something that makes you feel successful, helpful, etc. Example: "I am good with kids," "I'm a good listener," I am helpful and kind."
   - Talent: Something you can do well (can be born with or can be developed). In many cases, you either have it or you don't. Example: sports, playing a musical instrument, dance, etc.
2. Once students have generated their lists, they may either write or illustrate them on the Strengths & Talents sheet.
3. Have students write how these strengths or talents help them in difficult times or what they are able to do or accomplish because they have these strengths and talents.

Processing

1. Discuss how students have or might use these characteristics to bounce back from a challenging or difficult life event.
2. Help students identify times or events in their lives where these strengths or talents have helped them overcome difficulty. You may also refer to the first lifeline to discuss specific times they have used these strengths or talents to deal with things that have happened in their lives.
3. Discuss with students how having these strengths and talents helps them both in good and bad times.
4. Ask: "What kinds of things can you contribute to your community because you have these strengths and talents? How might you use these strengths and talents to help others?"

STRENGTHS & TALENTS

STRENGTHS	How does this strength help me?

TALENTS	How does this talent help me?

LESSON FOUR: YOU'RE NOT ALONE

Summary

Students will identify individual issues that are difficult for them but then work cooperatively with others to release or get rid of the stress associated with the problem. This is a fun activity that promotes teamwork but also shows that everyone has stress but it is possible to release the stress.

ASCA Mindsets & Behaviors

- M 3. Sense of belonging in the school environment
- B-LS 2. Demonstrate creativity
- B-SS 2. Create positive and supportive relationships with other students

Objectives

Students will:

- Understand everyone has stress and problems
- Demonstrate how to work cooperatively with others as a team
- Assess how to use problem-solving skills to work with others to achieve a goal

Materials

Balloons (one per person), masking tape for start and finish lines, permanent markers to write on balloons

Preparation

Inflate one balloon per person (extras in the event of popping). Make a start and finish line with masking tape.

Activity

1. Have students write on the balloons something related to their counseling goals. Examples: Things that are important to you, things that are necessary to do well in school, all the things you have to "keep up in the air," stressors you have to handle. When students have written on the balloon, discuss why what they have written is important or difficult.
2. Establish a finish line 50 yards away (depending on space). Explain that the task is to carry all these things to the next level/grade/stage of life or to carry these "burdens" to the finish line so they can let them go. They must carry all the balloons together (as a team) and must also stay connected together in some way. The balloons must all be transported by the group members while the group members are connected.
3. Rules: No use of hands or elbows to hold the balloons and no holding balloon in teeth. If anyone drops a balloon as the group is moving toward the finish line, the whole group must shout: "Oh no! We've dropped the (whatever is written on the balloon)!" and go back to the starting line.
4. Give the group a time limit, typically one–three minutes depending on age and frustration level of group.

Processing

1. What are some things you did not put on the balloon that might be important or difficult right now?
2. How did the group decide on a strategy to transport the balloons to the finish line? Would you do anything differently next time? If so, what?
3. How did it feel when someone dropped his/her balloon? Was it an individual problem or a group problem?
4. How was the process of making decisions about moving the balloons similar to the process you use to make decisions in situations at home or school?

Adapted from Kottman, T., Ashby, J.S., & DeGraaf, D. (2001)

LESSON 5: WHEN ONE DOOR CLOSES...THEN WHAT?

Summary

Students will discuss times when things have not turned out the way they planned and what happened instead. This activity helps students see that they can turn a negative into a positive.

ASCA Mindsets & Behaviors

- M 2. Self-confidence in ability to succeed
- B-SMS 6. Demonstrate ability to overcome barriers to learning
- B-SMS 7. Demonstrate effective coping skills when faced with a problem

Objectives

Students will:

- Understand how a negative event or action can have a positive result
- Apply the "doors" activity to their own lives and discuss positive ways to look at things

Materials

Doors handout, optical illusions website or handouts (e.g., *www.brainden.com*), list of successful people who failed before they succeeded (internet), glass of water filled halfway (optional)

Activity

1. Ask students if there has ever been a time when something bad happened to them or someone else. Give them time to discuss this event or circumstance. Ask how they felt about that event.
2. Show students an optical illusion (e.g., old woman/young woman) and ask what they see. Have students see if they can find both images in the picture. Discuss how people might see different things.
3. Define "perception." Perception is how we perceive or look at something. It may also be an impression or feeling that we have. Ask: Are our perceptions right or wrong? (If you are using the glass half empty/half full, talk about it. Is the glass half empty or half full?)
4. Discuss what it means to be an optimist or pessimist. Students can generate a definition or can describe how each sees or perceives life.

5. Ask students to remember the bad thing that happened to them or someone else. Ask: How did you "look" at the situation? What did you perceive about the event?" Then ask: Did things turn out the way that you thought they would?
6. Distribute the "Doors" activity. Ask students to identify at least two situations where one door closed, but another door opened. Process with students that even when something negative happens, they can usually find at least one positive thing.
7. Review a list of successful people who failed before they succeeded. Discuss why it is important to keep trying or to change their perspective when something does not go the way they would like.

Doors: It's How You Look at It!

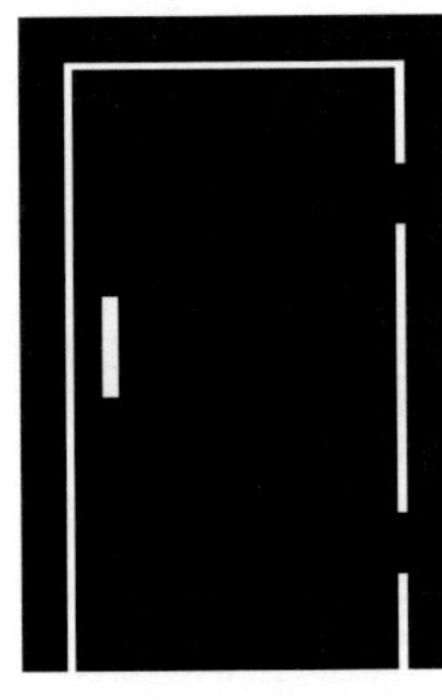

Door that closed ______________________________

____

Door that opened ______________________________

____

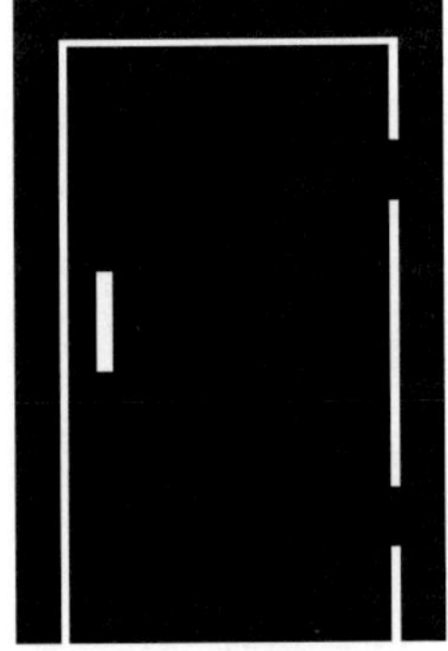

Door that closed ______________________________

____

Door that opened ______________________________

____

LESSON SIX: YOUR STORY IS UP TO YOU

Summary

Students will complete the lifeline activity but focus on the future. The goal is to have students then compare the future lifeline with the one completed in the first session and discover that what happens in their future becomes more in their control. The goal is for students to realize that despite the past, they have choices for their future.

ASCA Mindsets & Behaviors

- M 2. Self-confidence in ability to succeed
- B-LS 7. Identify long- and short-term academic, career and social/emotional goals
- B-SMS 5. Demonstrate perseverance to achieve long- and short-term goals

Objectives

Students will:

- Identify academic, college/career and social/emotional goals for their lives
- Understand consequences of decisions and choices
- Discover they have control over the decisions they make in their future

Materials

Copy of the lifeline (see Lesson 1), scratch paper, copies of the post-test on resilience

Activity

1. On a sheet of scratch paper, have students brainstorm what they would like to have happen in their lives from today until the "end of their story." Have students be specific (age, place, event, etc.). Ask students to transfer this information to the lifeline on the side labeled "Event."
2. Once students have listed their life events, go back and discuss how they would feel if that event occurred and who or what would control whether or not that event happens. The goal is to have the students see they have some control over what happens in their lives.
3. Have students list their feelings and control responses on the "Feelings/Control" side of the lifeline.

Processing

1. Discuss what the student will need to do to achieve the future event. Include the "reality" of events (such as becoming a professional basketball player) as well as the details about what it will take to make each event happen. Emphasize the importance of what students can do now to promote future success. For example, it will be difficult to go to college if students don't do well academically throughout school.
2. Compare this lifeline with the one the students did at the beginning of the program. Ask: "What do you notice about how things change on the side marked "Feelings/Control"? Help students understand that as they get older, they have more control over what happens to them.
3. Have students add the future lifelines to their lifebooks or folders.
4. Give students the post-test for resilience.

Evaluation

Compare pre- and post-test results to see if students have changed their perceptions. You may also want to check grades, attendance or behavior reports to see if there has been any change.

SECTION 2
STRENGTHS-BUILDING ACTIVITIES

These activities may be used in conjunction with the resilience activities, or they may be used separately. The pre-/post-test evaluation could also be used to collect process data about students' general feelings and attitudes regarding their hopefulness about the future. The activities are based on the signature strengths concepts of positive psychology by Dr. Martin Seligman (2002). They were developed with his permission.

Pre-/Post-test: Strengths

1. I feel comfortable with the way things are in the world.	Yes	No
2. I feel comfortable with the way things are in my community.	Yes	No
3. I feel comfortable with the way things are in my school.	Yes	No
4. I can cope when bad things happen.	Yes	No
5. I have people who support me when bad things happen.	Yes	No
6. I am hopeful about the future.	Yes	No
7. I believe I can achieve the goals in my life.	Yes	No

Three things I can do when bad things happen in the world:

____

____

____

Three things I can do when bad things happen to me:

____

____

____

LESSON ONE: THINGS MAY NOT BE WHAT THEY SEEM

Signature Strengths: Wisdom and knowledge (judgment/critical thinking/open-mindedness)

Summary

Students will play the "gift grab" game solely on their perceptions of what is inside the gift based on the outside packaging. Revealing what is inside each gift will help students understand the need to be careful before making a judgment about someone or something.

ASCA Mindsets & Behaviors

- B-LS 1. Demonstrate critical-thinking skills to make informed decisions
- B-SMS 2. Demonstrate self-discipline and self-control
- B-SS 9. Demonstrate social maturity and behaviors appropriate to the situation and environment

Objectives

Students will:

- Demonstrate positive behaviors in their interactions with others
- Predict what they think is in each gift
- Assess and conclude that what they often think or assume about something may not be reality

Materials

Gift boxes or bags (variety of shapes and sizes), some decorated, some not; the same gift in each box; slips of papers with numbers with enough for each of the participants

Activity

1. Place the boxes in the center of the group/room, and have participants draw a number to determine order of gift selection.
2. The person with number one goes first and chooses a package. Do not let the student open the gift.
3. Person with number two has the option of either taking person one's gift (unopened) or picking a new one from the pile. If he/she takes number one's gift, then number one gets to pick from the unchosen gifts.
4. Person number three has the option of taking person number one or two's gift or picking a new one from the pile. If anyone's gift gets taken, he/she has the choice of taking someone else's or picking from the pile.
5. When the gift has been stolen three times, the gift is "frozen" and cannot be selected again.
6. Continue until everyone has a gift.
7. Have students open their gifts.

Processing

1. Ask students why they chose the gift they did. Was it because of how it was decorated? Did they figure the prettiest or largest gift would be the best?
2. Ask: "Which boxes/bags had the best gifts?" "Could you tell from the outside which gifts would be the best on the inside?"
3. Ask: "How did it feel when someone took your gift, even though you didn't really know what the gift was?"
4. Ask: "How might you have played differently if you knew what was inside?"
5. Ask: "What does this tell us about making judgments or decisions based on how things look or seem on the outside?" (Don't make assumptions, don't always think things are the way they look, etc.)
6. When bad things happen and we see or hear about it from TV or somewhere else (could discuss sources of information), how should we interpret what has happened? (You may want to discuss how the media sensationalizes things, focuses on the drama, etc.).

LESSON TWO: WHAT YOU FEEL IS REAL

Signature Strengths: Courage (integrity/genuineness/honesty) and temperance (self-control)

Summary

Students will identify and acknowledge when they have strong feelings about something. The purpose of this activity is to validate students' true feelings but also consider a new or reframed feeling that might lead to more positive results.

ASCA Mindsets & Behaviors

- M 1. Belief in development of whole self, including a healthy balance of mental, social/emotional and physical well-being
- B-LS 9. Gather evidence and consider multiple perspectives to make informed decisions
- B-SMS 2. Demonstrate self-discipline and responsibility

Objectives

Students will:

- Identify and express feelings
- Learn coping skills for managing life events
- Evaluate how reframing a situation may lead to more positive outcome

Materials

What you Feel is Real handout, song "True Colors" (Cyndi Lauper or Phil Collins) and lyrics sheet. Song lyrics can be downloaded from the Internet.

Activity

1. Distribute handout to students. Talk about things that have happened in the world recently. Have them list the event(s) and how they really felt about it.
2. In the "Reframe" column, have students come up with a positive way to respond to their feelings (whether their feelings were positive or negative).

3. After doing the "global" list, have students list the events that have happened in their own lives and how they felt about them.
4. In the "Reframe" column, have students come up with a positive way to respond to their feelings.
5. Distribute the lyrics to "True Colors," and have students listen to the song.

Processing

1. Validate students' feelings as they list them. It is OK to acknowledge your true feelings ("true colors") about something.
2. Discuss the consequences of reacting to initial feelings, which are often negative or destructive.
3. Talk about power. Ask: "Who has the power if you act on those feelings? Who has the power to change those feelings so they are expressed in a positive way?"
4. Ask: "How does it help if we respond with a reframed feeling that is more positive?" Discuss the consequences of more constructive expressions of emotions.
5. Ask: "Why is it important to have power or control over your emotions? Has there been a time in your life where expressing your emotions in a positive way helped you or someone else?" Have students share.

What You Feel is Real: Be True to You

Event/What happened?	How did you really feel?	Reframed or positive feeling
1.		
2.		
3.		
4.		
5.		
6.		
7.		

Consequences of acting on first feelings:

___

___

___

Consequences of acting on reframed feelings:

___

___

___

LESSON THREE: MY IGLOO OF INFLUENCE AND INSPIRATION

Signature Strength: Justice (citizenship/teamwork/loyalty) and humanity and love (loving and allowing oneself to be loved)

Summary

Students will identify the people in their lives who are resources for them in times of adversity. By concretely naming people who accept and support them, students will feel connected to others and understand the importance of reaching out for help when needed.

ASCA Mindsets & Behaviors

- M 2. Self-confidence in ability to succeed
- B-SMS 7. Demonstrate effective coping skills when faced with a problem
- B-SS 2. Create positive and supportive relationships with other students
- B-SS 3. Create relationships with adults that support success

Objectives

Students will:

- Identify resource people in the school and community and know how to seek their help
- Choose people in their lives and explain how they help or support them in different ways

Material

Igloo of Influence handout, pen or pencil

Activity

1. Distribute the handout and discuss what it means to influence someone.
2. Discuss the difference between someone who is important and someone who is influential.
3. Discuss how ice is often seen as cold, but when blocks of ice are put together, they surround you with warmth and safety.
4. Have them write names for each of the questions: Who inspires you? Who comforts you? Who energizes you (makes them want to do better)?
5. Students may also want to write the names of people who support them on the blocks of ice on the igloo.

Processing

1. Have students share the names they have written and describe how each person inspires, comforts or energizes them.
2. Ask: "Under what circumstances/situations would you turn to someone on the list?"
3. Ask: "Would you be one of the blocks of ice in someone else's igloo? Who would say you are someone who inspires, comforts or energizes them?"
4. Ask participants to keep a copy of this sheet accessible. Make a copy to keep in the school counseling office in the event a student needs to be reminded of the human resources available to him/her.

Igloo of Influence and Inspiration

Igloos are made of ice blocks. Although we think of ice as cold, we know that when the blocks of ice are connected together, they provide shelter, warmth and protection.

Think of the people who help protect and influence you. Write their names on the blocks of ice. Then fill in the blanks below:

Who influences you? ____

____

Who comforts you? ____

____

Who energizes you? ____

____

Even when the world seems cold, you can surround yourself with people who will comfort and strengthen you.

List people who might put you as a block in their igloo. Who do you help or support?

____

____

LESSON FOUR: THE DEAL OF YOUR LIFE

Signature Strength: Temperance (self-control, priorities)

Summary

This game will help students identify what is important to them in their life. It also opens up the discussion of "playing the hand you are dealt" and thinking of ways to change things if they need to be changed.

ASCA Mindsets & Behaviors

- M 6. Positive attitude toward work and learning
- B-LS 1. Demonstrate critical-thinking skills to make informed decisions
- B-SMS 7. Demonstrate effective coping skills when faced with a problem

Objectives

Students will:

- Evaluate how school success and academic achievement enhance future career and vocational opportunities
- Compare and contrast the impact of their choices and priorities

Materials

Deck of cards, enough copies of Symbol Sheet for each member, one sheet of blank paper for each participant

Activity

1. Use one deck of cards. If playing with three or fewer people, deal 15 cards each; if more, then deal the whole deck. The value of the card does not matter, just the suit.
2. Have participants group according to suit; students will rank order the importance of each suit.
3. Participants will take turns laying down their cards and discuss if they have a balanced life with the following questions.
   - Were you dealt a balanced life?
   - What cards would help you increase the balance of your life?
   - What cards are not helping you?
4. As participants share their hands, have them determine the positive and negative factors associated with "the hand they were dealt." What is missing? What is there too much of? What are the consequences of the imbalance?
5. After discussion, allow participants to talk about how they might help each other achieve more balance; allow cards to be shared at this point.

Processing

1. Ask: "Did someone else get the life you wanted? Why or why not?"
2. Ask: "If this were really your life, how would you handle it? What would you do to change it?"
3. Discuss that life may deal them a bad hand, and it is up to them how they play that hand. What are their options?
4. Discuss what we do and don't have control over in our lives (e.g., we may have control over our income, but maybe we cannot afford to college).

Lesson Source: Starr, A.F. (2006). Group counseling course curriculum. Arlington, VA: Marymount University

The Deal of Your Life Symbol Sheet

Diamonds = Money Income
Clubs = Social Life/Friends
Spades = Education/School
Heart = Family/Support

Source: Amy Starr (2006). *Group counseling curriculum project.* Arlington, VA: Marymount University:

LESSON FIVE: EXPRESSING GRATITUDE

Signature Strength: Transcendence (gratitude)

Summary

Students will discuss the meaning of gratitude and apply it to their own lives. The activity will provide a way for them to express gratitude to someone in their lives.

ASCA Mindsets & Behaviors

- M 6. Positive attitude toward work and learning
- B-SS 1. Use effective oral and written communication skills and listening skills
- B-SS 9. Demonstrate social maturity and behaviors appropriate to the situation and environment

Objectives

Students will:

- List things and people for which they are grateful
- Express gratitude through writing
- Practice gratitude through delivering the card to the intended person
- Value the experience of expressing gratitude

Materials

Blank thank you cards or large unlined index cards, pens or markers

Activity

1. Discuss what it means to be thankful or grateful. Show a YouTube video as an example of what gratitude means. Generate a list of things students are thankful for.
2. Discuss the difference between being grateful for things and being grateful for people.
3. Talk about how it feels to have someone say thank you, and ask students to count how many times they have said thank you that day.
4. Talk about how it feels to say thank you, and have students think about someone in their lives who deserves a thank you.
5. Distribute the blank thank you cards, and ask students to write or draw a card to someone they are grateful to (could be for something specific or general).
6. Ask students to share.
7. Ask students to send or give the card to the person they need to thank.

Processing

1. Discuss why people do not say thank you very much.
2. Talk about situations where we can say thank you, even in situations where we might not know the person (e.g., store, school, phone).
3. Talk about why it is important to say thank you to people.
4. Ask students to try to be more grateful and thankful to people around them.

LESSON SIX: THE POWER OF FORGIVENESS

Signature Strength: Transcendence (forgiveness and mercy)

Summary

Students will learn forgiveness is more for the forgiver than the person to be forgiven. The activities will help the students literally feel the release of a heavy load and understand that forgiveness can lead to inner peace.

ASCA Mindsets & Behaviors

- M 1. Belief in development of whole self, including a healthy balance of mental, social/emotional and physical well-being
- B-SMS 2. Demonstrate self-discipline and self-control
- B-SS 8. Demonstrate advocacy skills and ability to assert self, when necessary

Objectives

Students will:

- Accept mistakes as essential to the learning process
- Respect alternative points of view
- Develop effective coping skills for dealing with problems

Materials

Forgiveness stone (one per student), a big cloth bag, bricks or books

Activity

1. Talk about forgiveness: How willing are participants to forgive someone who does something bad (either to them or to others)?
2. Ask if anyone can tell a story about a time he/she has forgiven someone and how he/she felt.
3. Ask a student to tell a story about when someone did something wrong to him/her.
4. Have the student name the feelings or bad things associated with that experience.
5. For each feeling or bad thing the student names, add a brick or book to the bag. Have the student pick up the bag after each brick and talk about how heavy the bag becomes with each brick (or bad feeling).
6. Modification: Have students hold a textbook in front of them at shoulder level. The longer they hold it, the heavier the book becomes. Talk about the options (let it go, drop it or continue to hold as it becomes more uncomfortable) and the consequences of each choice.
7. Give each participant a forgiveness stone (may want to call it a "sorry stone"). Students may want to paint or decorate the stone as a reminder to forgive others.

Processing:

1. Ask: "How does it feel to carry around feelings of anger or revenge? Who does it hurt?"
2. Discuss: "Some people feel forgiveness allows them to let go of negative feelings so they can get on with their lives. What do you think about that?"
3. Ask participants to keep the stone to remind them that forgiveness will help them more than anyone else.

LESSON SEVEN: HOPES AND HURDLES

Signature Strengths: Courage (persistence) and transcendence (hope)

Summary

Students will plan for their future. This session allows students to identify where they see themselves in the future and what skills, attitudes and resources they may need to achieve their goal. Students will also consider what hurdles or challenges might get in the way of realizing their future hopes and discuss ways they might confront those difficulties to reach their goal.

ASCA Mindsets & Behaviors

- M 5. Belief in using abilities to their fullest to achieve high-quality results and outcomes
- B-LS 9. Gather evidence and consider multiple perspectives to make informed decisions
- B-SMS 5. Demonstrate perseverance to achieve long- and short-term goals

Objectives

Students will:

- Identify goals for the future
- Analyze hurdles/challenges and how to overcome them

Materials

Hopes and Hurdles handout, pencils or pens, lemons or lemonade

Activity

1. Ask students to think about themselves in the next five-10 years (For younger students, make it enough years for them to be out of high school.)
2. Give each participant a lemon. Discuss the saying, "When life gives you lemons, make lemonade."
3. Distribute the Hopes and Hurdles handout. Have students complete the handout.
4. Have students share their responses to their handout and discuss. Have other students offer ways to overcome challenges to each other's goals.
5. On the scaling question (next to last item on handout), ask students how they scored themselves on the likelihood of getting what they hope for. Ask: What would help increase your score by one number? If students have ranked themselves a 10, ask: What might happen that would decrease your number, and how might you handle it?"
6. End the session with lemonade. Share what advice each student would give themselves in 10 years.
7. Distribute the post-test survey from the beginning session.

Hopes and Hurdles

Where do you see yourself in five–10 Years? ______________________________

What skills will you need? ______________________________

What attitude will you need? ______________________________

What resources will you need? ______________________________

Who will help you get there? ______________________________

What challenges will you face? ______________________________

Who or what might stand in your way? ______________________________

How will you overcome those challenges that get in your way? ______________________________

____

On a scale of 1 to 10, with 1 being Totally Unlikely and 10 being Totally Likely, how likely are you to realize your hopes and reach your goals?

1 2 3 4 5 6 7 8 9 10

Imagine yourself in 10 years. How old will you be? _______ What positive message would you say to younger self? ______________________________

____

____

CONCLUSION

After completing these activities, it will be important to check-in and follow-up with students on a regular basis. Although completing the activities will help students understand they can handle difficult challenges, it is important to reinforce the concepts emphasized in these sessions. Reminding students of their internal and external assets will help students realize the resources they can access for support and assistance when things go wrong in their own lives or in the world.

It may also be beneficial to use these activities in other venues, perhaps with other school faculty or staff. We should not presume that life circumstances and world events only affect our youth. We should remind ourselves that despite seemingly hopeless situations and catastrophic events, we can be resilient, and we can all find ways to move forward with optimism and hope.

ASCA MINDSETS & BEHAVIORS FOR STUDENT SUCCESS

The ASCA Mindsets & Behaviors for Student Success: K-12 College- and Career Readiness for Every Student describe the knowledge, skills and attitudes students need to achieve academic success, college and career readiness and social/emotional development. The standards are based on a survey of research and best practices in student achievement from a wide array of educational standards and efforts. These standards are the next generation of the ASCA National Standards for Students, which were first published in 1997.

The 35 mindset and behavior standards identify and prioritize the specific attitudes, knowledge and skills students should be able to demonstrate as a result of a school counseling program. School counselors use the standards to assess student growth and development, guide the development of strategies and activities and create a program that helps students achieve their highest potential. The ASCA Mindsets & Behaviors can be aligned with initiatives at the district, state and national to reflect the district's local priorities.

To operationalize the standards, school counselors select competencies that align with the specific standards and become the foundation for classroom lessons, small groups and activities addressing student developmental needs. The competencies directly reflect the vision, mission and goals of the comprehensive school counseling program and align with the school's academic mission.

RESEARCH-BASED STANDARDS

The ASCA Mindsets & Behaviors are based on a review of research and college- and career-readiness documents created by a variety of organizations that have identified strategies making an impact on student achievement and academic performance. The ASCA Mindsets & Behaviors are organized based on the framework of noncognitive factors presented in the critical literature review "Teaching Adolescents to Become Learners" conducted by the University of Chicago Consortium on Chicago School Research (2012).

This literature review recognizes that content knowledge and academic skills are only part of the equation for student success. "School performance is a complex phenomenon, shaped by a wide variety of factors intrinsic to students and the external environment" (University of Chicago, 2012, p. 2). The ASCA Mindsets & Behaviors are based on the evidence of the importance of these factors.

ORGANIZATION OF THE ASCA MINDSETS & BEHAVIORS

The ASCA Mindsets & Behaviors are organized by domains, standards arranged within categories and subcategories and grade-level competencies. Each is described below.

DOMAINS

The ASCA Mindsets & Behaviors are organized in three broad domains: academic, career and social/emotional development. These domains promote mindsets and behaviors that enhance the learning process and create a culture of college and career readiness for all students. The definitions of each domain are as follows:

- **Academic Development** – Standards guiding school counseling programs to implement strategies and activities to support and maximize each student's ability to learn.
- **Career Development** – Standards guiding school counseling programs to help students 1) understand the connection between school and the world of work and 2) plan for and make a successful transition from school to postsecondary education and/or the world of work and from job to job across the life span.
- **Social/Emotional Development** – Standards guiding school counseling programs to help students manage emotions and learn and apply interpersonal skills.

STANDARDS

All 35 standards can be applied to any of the three domains, and the school counselor selects a domain and standard based on the needs of the school, classroom, small group or individual. The standards are arranged within categories and subcategories based on five general categories of noncognitive factors related to academic performance as identified in the 2012 literature review published by the University of Chicago Consortium on Chicago School Research. These categories synthesize the "vast array of research literature" (p. 8) on noncognitive factors including persistence, resilience, grit, goal-setting, help-seeking, cooperation, conscientiousness, self-efficacy, self-regulation, self-control, self-discipline, motivation, mindsets, effort, work habits, organization, homework completion, learning strategies and study skills, among others.

- **Category 1: Mindset Standards** – Includes standards related to the psycho-social attitudes or beliefs students have about themselves in relation to academic work. These make up the students' belief system as exhibited in behaviors.
- **Category 2: Behavior Standards** – These standards include behaviors commonly associated with being a successful student. These behaviors are visible, outward signs that a student is engaged and putting forth effort to learn. The behaviors are grouped into three subcategories.
  a. **Learning Strategies:** Processes and tactics students employ to aid in the cognitive work of thinking, remembering or learning.
  b. **Self-management Skills:** Continued focus on a goal despite obstacles (grit or persistence) and avoidance of distractions or temptations to prioritize higher pursuits over lower pleasures (delayed gratification, self-discipline, self-control).
  c. **Social Skills:** Acceptable behaviors that improve social interactions, such as those between peers or between students and adults.

THE ASCA MINDSETS & BEHAVIORS FOR STUDENT SUCCESS: K-12 COLLEGE- AND CAREER-READINESS STANDARDS FOR EVERY STUDENT

Each of the following standards can be applied to the academic, career and social/emotional domains.

CATEGORY 1: MINDSET STANDARDS

School counselors encourage the following mindsets for all students.

M 1. Belief in development of whole self, including a healthy balance of mental, social/emotional and physical well-being

M 2. Self-confidence in ability to succeed

M 3. Sense of belonging in the school environment

M 4. Understanding that postsecondary education and life-long learning are necessary for long-term career success

M 5. Belief in using abilities to their fullest to achieve high-quality results and outcomes

M 6. Positive attitude toward work and learning

CATEGORY 2: BEHAVIOR STANDARDS

Students will demonstrate the following standards through classroom lessons, activities and/or individual/small-group counseling.

Learning Strategies	Self-Management Skills	Social Skills
B-LS 1. Demonstrate critical-thinking skills to make informed decisions	**B-SMS 1.** Demonstrate ability to assume responsibility	**B-SS 1.** Use effective oral and written communication skills and listening skills
B-LS 2. Demonstrate creativity	**B-SMS 2.** Demonstrate self-discipline and self-control	**B-SS 2.** Create positive and supportive relationships with other students
B-LS 3. Use time-management, organizational and study skills	**B-SMS 3.** Demonstrate ability to work independently	**B-SS 3.** Create relationships with adults that support success
B-LS 4. Apply self-motivation and self-direction to learning	**B-SMS 4.** Demonstrate ability to delay immediate gratification for long-term rewards	**B-SS 4.** Demonstrate empathy
B-LS 5. Apply media and technology skills	**B-SMS 5.** Demonstrate perseverance to achieve long- and short-term goals	**B-SS 5.** Demonstrate ethical decision-making and social responsibility
B-LS 6. Set high standards of quality	**B-SMS 6.** Demonstrate ability to overcome barriers to learning	**B-SS 6.** Use effective collaboration and cooperation skills
B-LS 7. Identify long- and short-term academic, career and social/emotional goals	**B-SMS 7.** Demonstrate effective coping skills when faced with a problem	**B-SS 7.** Use leadership and teamwork skills to work effectively in diverse teams
B-LS 8. Actively engage in challenging coursework	**B-SMS 8.** Demonstrate the ability to balance school, home and community activities	**B-SS 8.** Demonstrate advocacy skills and ability to assert self, when necessary
B-LS 9. Gather evidence and consider multiple perspectives to make informed decisions	**B-SMS 9.** Demonstrate personal safety skills	**B-SS 9.** Demonstrate social maturity and behaviors appropriate to the situation and environment
B-LS 10. Participate in enrichment and extracurricular activities	**B-SMS 10.** Demonstrate ability to manage transitions and ability to adapt to changing situations and responsibilities	

GRADE-LEVEL COMPETENCIES

Grade-level competencies are specific, measurable expectations that students attain as they make progress toward the standards. As the school counseling program's vision, mission and program goals are aligned with the school's academic mission, school counseling standards and competencies are also aligned with academic content standards at the state and district level.

ASCA Mindsets & Behaviors align with specific standards from the Common Core State Standards through connections at the competency level. This alignment allows school counselors the opportunity to help students meet these college- and career-readiness standards in collaboration with academic content taught in core areas in the classroom. It also helps school counselors directly align with academic instruction when providing individual and small-group counseling by focusing on standards and competencies addressing a student's developmental needs. School counselors working in states that have not adopted the Common Core State Standards are encouraged to align competencies with their state's academic standards and can use the competencies from the ASCA Mindsets & Behaviors as examples of alignment.

ASCA MINDSETS & BEHAVIORS DATABASE

The grade-level competencies are housed in the ASCA Mindsets & Behaviors database at *www.schoolcounselor.org/studentcompetencies*. School counselors can search the database by keyword to quickly and easily identify competencies that will meet student developmental needs and align with academic content as appropriate. The database also allows school counselors to contribute to the competencies by sharing other ways to meet or align with a specific standard.

REFERENCES AND RECOMMENDATIONS

American School Counselor Association (2014). *Mindsets and Behaviors for Student Success: K-12 College- and Career-Readiness Standards for Every Student.* Alexandria, VA: Author.

Bosworth, K. & Walz, G.R. (2005) *Promoting student resiliency.* Alexandria, VA: American Counseling Association

Davis, T.E. (1997). Telling life stories and creating lifebooks: A counseling technique for fostering resilience in children. *Dissertation Abstracts International, 58*(11A). Retrieved from https://vtechworks.lib.vt.edu/handle/10919/29178

Kottman, T., Ashby, J.S., & DeGraaf, D. (2001). *Adventures in guidance: How to integrate fun into your guidance program.* Alexandria, VA: American Counseling Association.

Masten, A. S., Cutuli, J. J., Herbers, J. E., & Reed, M.-G. J. (2009). Resilience in development. In C. R. Snyder & S. J. Lopez (Eds.), *Oxford handbook of positive psychology*, 2nd ed. (pp. 117 - 131). New York: Oxford University Press.

Seligman, M.E.P. (2002). *Authentic happiness: Using the new positive psychology to realize your potential for lasting fulfillment.* New York: Free Press.

Starr, A.F. (2006). *The deal of your life.* Group counseling curriculum project. Arlington, VA: Marymount University.

Website on positive psychology: *www.authentichappiness.org*